THE SOLAR SYSTEM FOR THE SMART KIDS

SIMPLE FACTS FOR SUPER KIDS

EMMY A.M

Copyright

Copyright © 2018 by Emmy A.M

All rights reserved. This book or any portion thereof may not be reproduced or used in any manner whatsoever without the express written permission of the publisher except for the use of brief quotations in a book review or scholarly journal.

First edition: 2018

DEDICATION

TO MY....

Amazing girl

Reliable friend

Warm heart

A delicious Cat

ARWA

WRITE YOUR DEDICATION HERE

WHAT'S IN OUR SOLAR SYSTEM?

*Our Solar System consists of a central star (the Sun), the nine planets orbiting the sun, moons, asteroids, comets, meteors, interplanetary gas, dust, and all the "space" in between them.

*The nine planets of the Solar System are named for Greek and Roman Gods and Goddesses.

INNER AND OUTER PLANETS

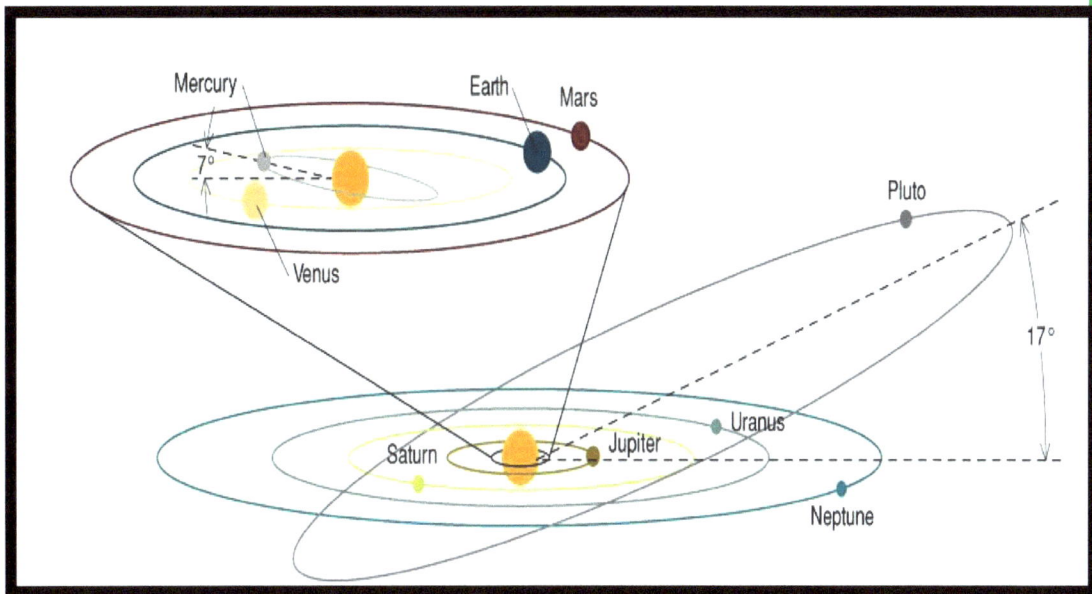

* **Inner Planets:**
  - Mercury
  - Venus
  - Earth
  - Mars

* **Outer Planets:**
  - Jupiter
  - Saturn
  - Uranus
  - Neptune
  - Pluto

THE SUN

- The sun's energy comes from nuclear fusion (where hydrogen is converted to helium) within its core.
- This energy is released from the sun in the form of heat and light.

- The Sun is a star at the center of our solar system.

- Stars produce light.
- Planets reflect light.

- A star's temperature determines its "color."
- The coldest stars are red.
- The hottest stars are blue.

THE SUN

- It supports all life on Earth through photo-synthesis and is the ultimate source of all food and fossil fuel.

- 99.86% of all the mass of the solar system is found in the Sun.

- The core of the Sun is 16 million °C.

- The surface of the Sun is 7000° C.

- The Sun generates energy the equivalent of 100 billion tons of TNT exploding every second.

NUCLEAR FUSION IN THE SUN

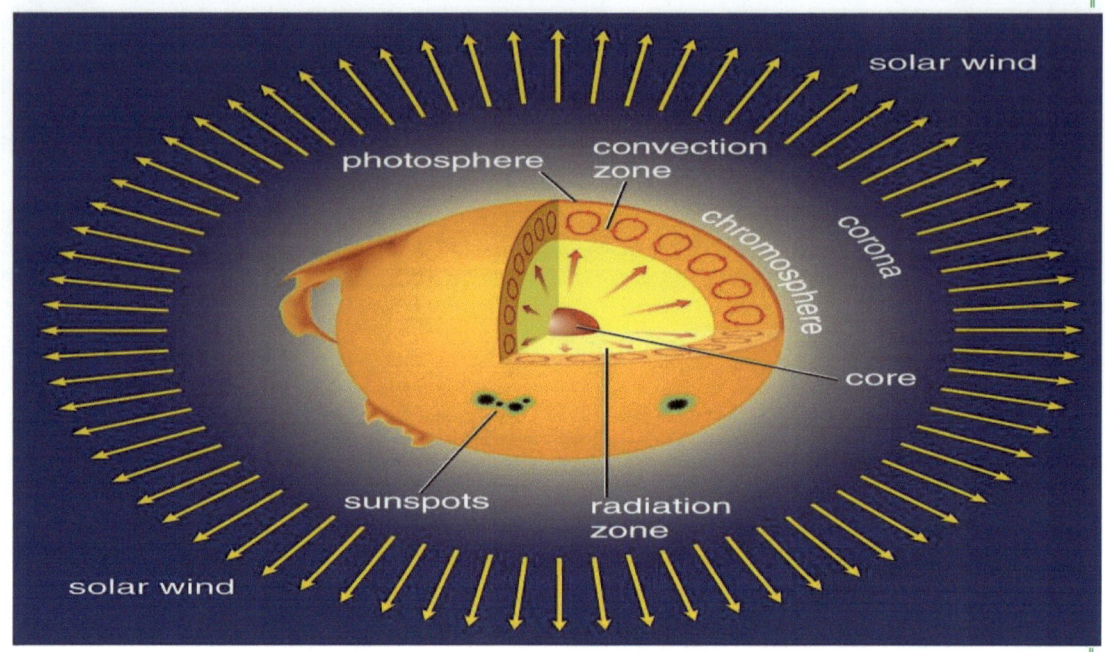

- The Sun is basically a huge ball of hydrogen gas held together by the gravity created by its own mass.

- Under the intense pressure created at the centre of the Sun by gravity, hydrogen nuclei are fused together to produce helium nuclei.

- Very simply stated, four hydrogen nuclei are fused into one helium nucleus, however one helium atom has less mass than four hydrogen atoms.

- The fusion process releases enough energy to account for the lost mass.

THE 9 PLANETS OF THE SOLAR SYSTEM

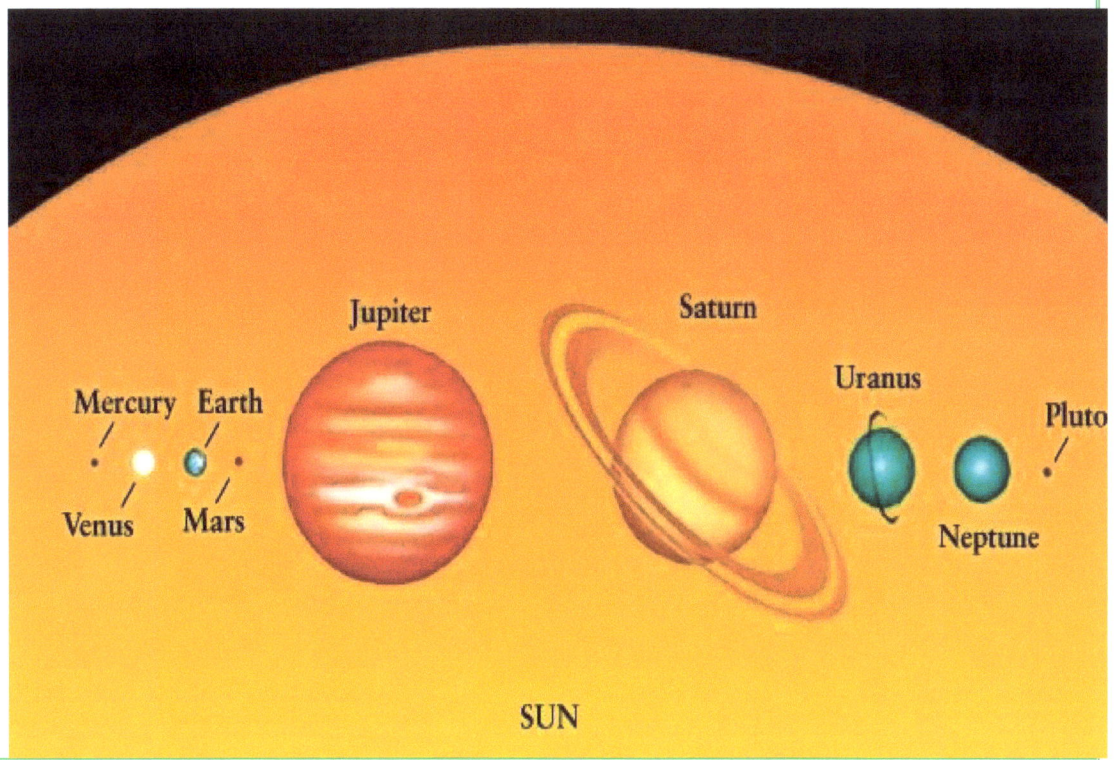

- Planets are categorized according to composition and size.

- There are two main categories of planets.

- Small rocky planets (Mercury, Venus, Earth, Mars, and Pluto).

- Gas giants (Jupiter, Saturn, Uranus, and Neptune).

CHARACTERISTICS OF SMALL ROCKY PLANETS

* The four rocky planets are Mercury, Venus, Earth and Mars.

* They are made up mostly of rock and metal.

* They are very heavy.

* They move slowly in space.

* They have no rings and few moons (if any).

* They have a diameter of less than 13,000 km.

MERCURY

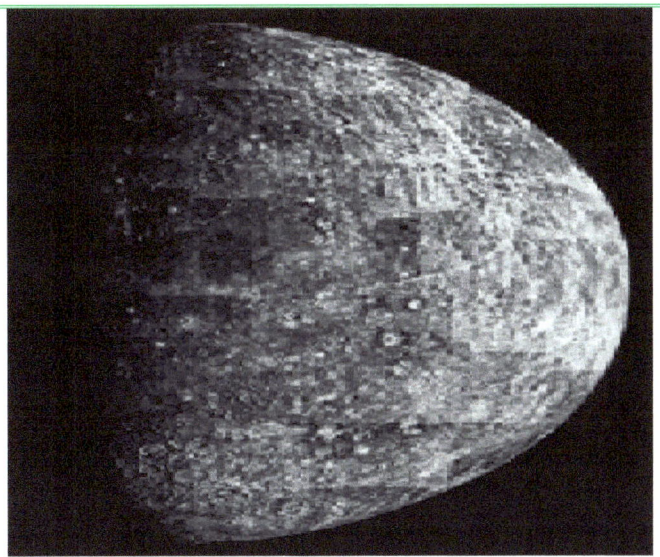

- Relative position: *1st planet out from the sun.*

- Scientists believe there is ICE on Mercury! The ice is protected from the sun's heat by crater shadows.

- Mercury has extreme temperature fluctuations, ranging from 800°F (daytime) to −270°F (nighttime).

- Temperatures: High: 467 °C on the sunny side of the planet. Low: −183 °C on the dark side of the planet

MERCURY

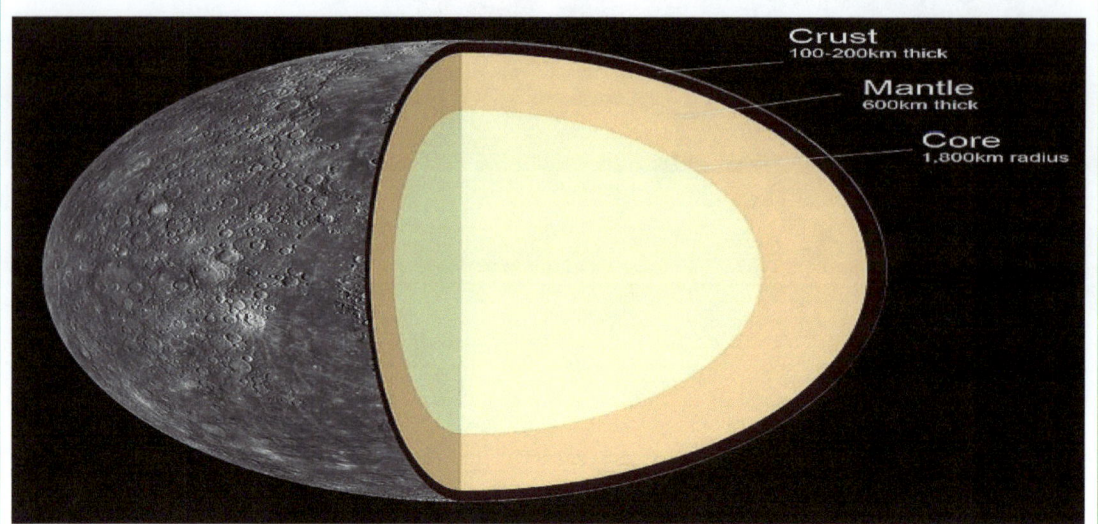

Internal structure of Mercury

•Atmosphere: Almost no atmosphere. •The very little atmosphere that exists is composed chiefly of oxygen, sodium, and helium.	(Moons): None.
	•Approximately half the size of Earth.
•Rotation: 58.65 days (very slow rotation)	•Mercury has a revolution period of 88 days.

VENUS

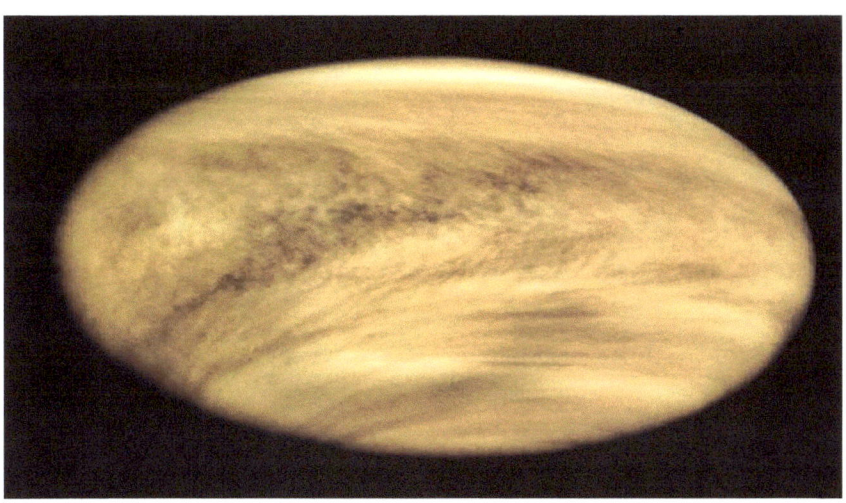

- Second planet from the sun.

- People often mistake it for a star. Venus is the brightest object in the sky after the sun and moon because its atmosphere reflects sunlight so well.

- Its maximum surface temperature may reach 900°F.

- Venus has no moons.

- Nicknamed "the evening star"
- Orange in Color

- Venus takes 225 days to complete an orbit.

VENUS

Internal Structure of Venus

- Crust: silicate rocks, unknown composition
- Rocky Mantle: unknown composition
- Atmosphere: 9.3MPa, mostly carbon dioxide, sulfuric acid cloud deck
- Metallic Core: partially solid, primarily iron/nickel

*Approximately the same size as the Earth.

*General composition: Rocky material. It contains an iron core and a molten rocky mantle. The crust is a solid, rocky material.

*It is a terrestrial planet.

*Atmosphere: Consists mainly of carbon dioxide, nitrogen, and droplets of sulfuric acid;

*it contains almost no water vapor.

*This thick atmosphere traps immense amounts of heat in a large-scale.

EARTH

- **Third** planet from the sun
- Has **white clouds**.
- **Earth** is the only planet **known to support living organisms**.
- **Earth's** surface is composed of **71% water**:
  1 - **Water** is necessary for life on Earth.
  2 - **The oceans** help maintain Earth's stable temperatures.
- **Earth** has **one moon** and an oxygen rich atmosphere.

EARTH

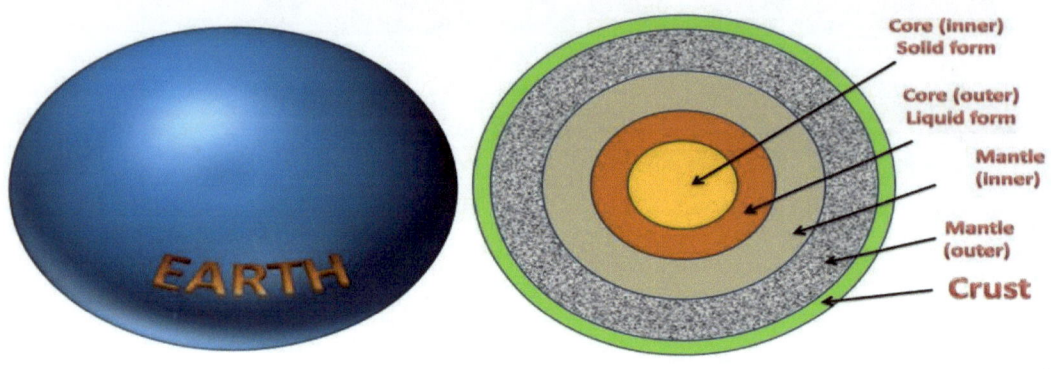

*General composition: Rocky material.

*It is a terrestrial planet. It has a nickel-iron core with a molten mantle and solid rocky crust.

*Rotation: 23 hours, 56 miutes (1 day).

*Appearance: The Earth looks blue and green from space with clouds moving through the atmosphere.

The surface of the Earth is 70% water and 30% land.

EARTH'S MOON

- It takes the moon approximately 29 days to complete one rotation.

- The same side of the moon always faces us.

- The moon's surface is covered in dust and rocky debris from meteor impacts.

- Earth's Moon has no water or atmosphere.

- The moon reflects light from the sun onto the earth's surface.

EARTH'S MOON

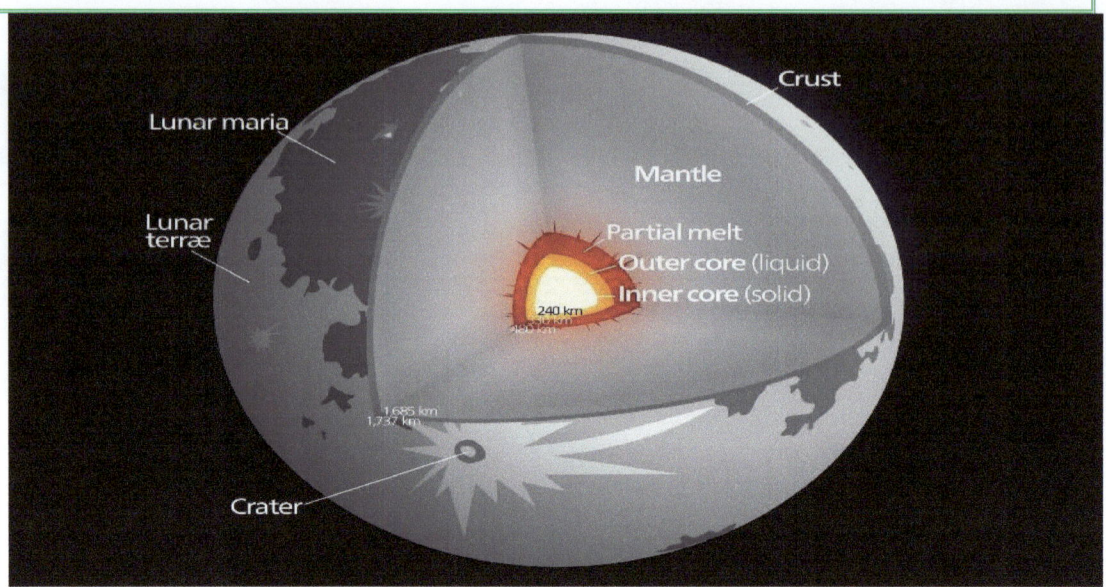

The Moon: Internal Structure

- It is 384 401 km from the Earth.

- The surface of the Moon is covered with craters and flatlands.

- The craters are due to repeated meteorite bombardments while the dark, flatlands are the result of ancient lava flows

- It has about 1/6 the mass of the Earth, therefore it has 1/6 the gravitational pull of the Earth.

MARS

- **4th** planet out from the sun.
- **Mars** Has **2 Moons**
- **Mars** appears red because of **iron oxide**, or rust, in its soil.
- **Mars** is Pink-Gold in color
- **Mars** has two moons and takes about **two years** to complete an **orbit**.
- **Mars** is The **RED** Planet

MARS

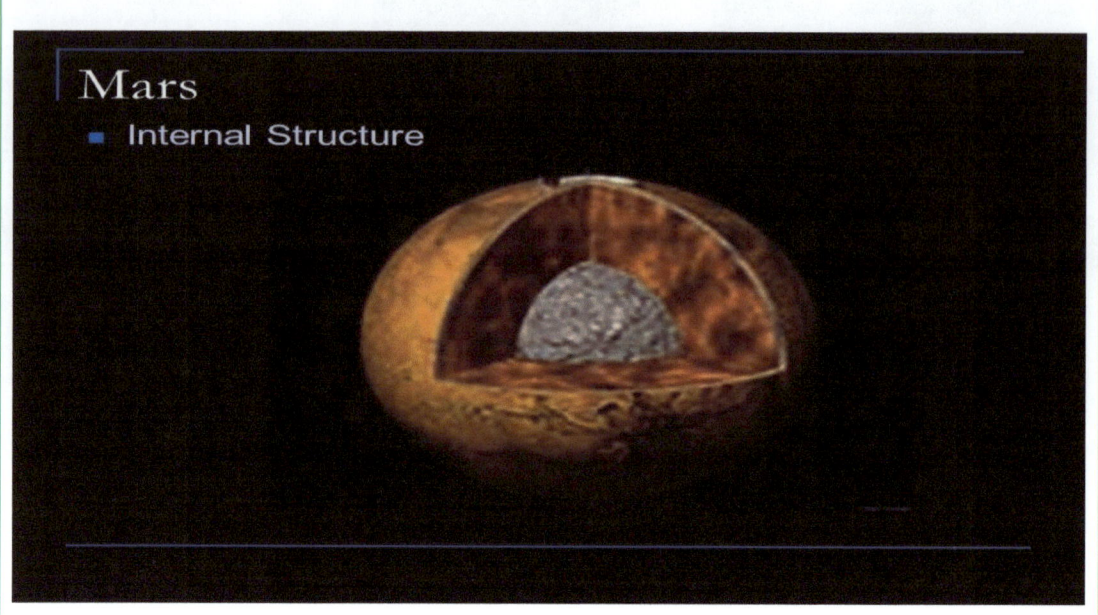

- Like Earth, Mars has ice caps at its poles.
- Mars appears red due to the iron oxide in its soil.
- Atmosphere: Mostly carbon dioxide, nitrogen, and argon.
- Mars has the largest volcanic mountain in the solar system.
- Mars is Approximately half the size of Earth
- Mars has four seasons similar to Earth.

MOONS OF MARS

Phobos:
* Gouged by a giant impact crater and beaten by thousands of meteorite impacts, Phobos is on a collision course with Mars.
* It may collide with Mars in 50 million years or break up into a ring.

Deimos:
* It is also heavily cratered with a small lumpy appearance.

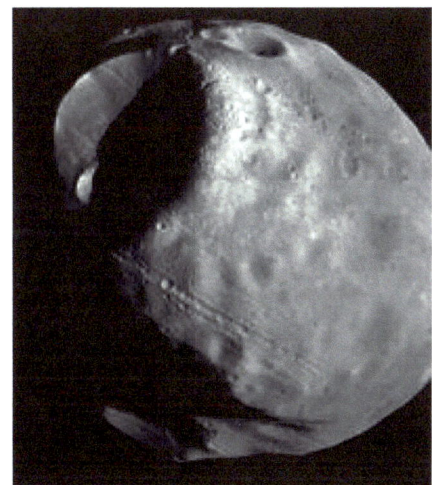

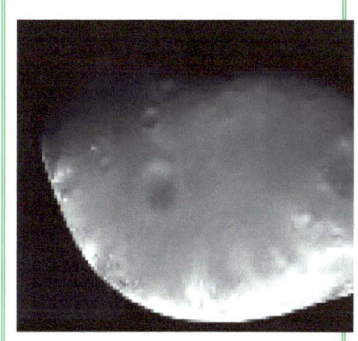

PLUTO

Pluto in Colorized Infrared

- Pluto has only one moon.

- Pluto takes about 249 years to orbit the sun.

- Part of Pluto's orbit passes inside that of Neptune, so at times Neptune is the planet farthest from the sun.

- Pluto was located and named in 1930

- Today Pluto is no longer considered a planet.

CHARACTERISTICS OF GAS GIANTS

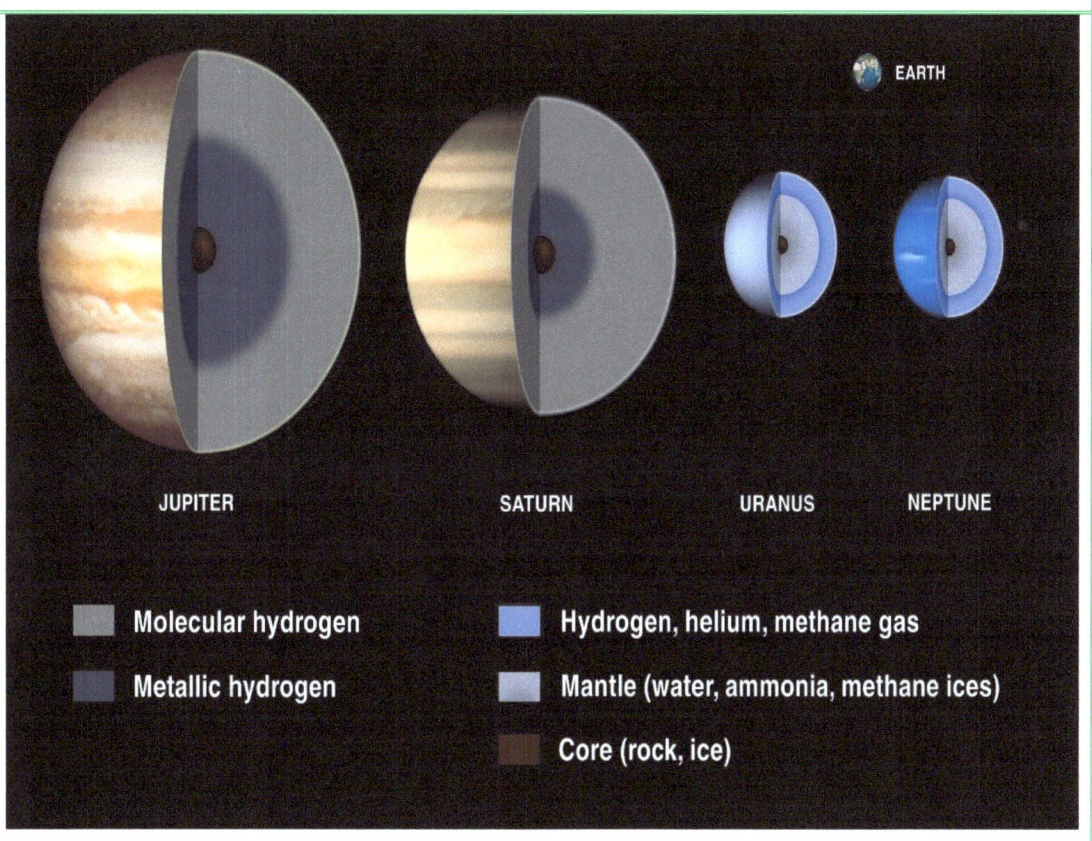

- They are made up mostly of gases (primarily hydrogen & helium).
- They are very light for their size.
- They move quickly in space.
- They have rings and many moons.
- They have a diameter of less than 48,000 km

JUPITER

- 5th planet out from the sun.
- It takes about 12 years for Jupiter to orbit the sun.
- Giant ball of swirling gas.
- Jupiter is largest Planet.
- Jupiter has 16 moons.
- Has a large red spot, which is a giant storm.

JUPITER

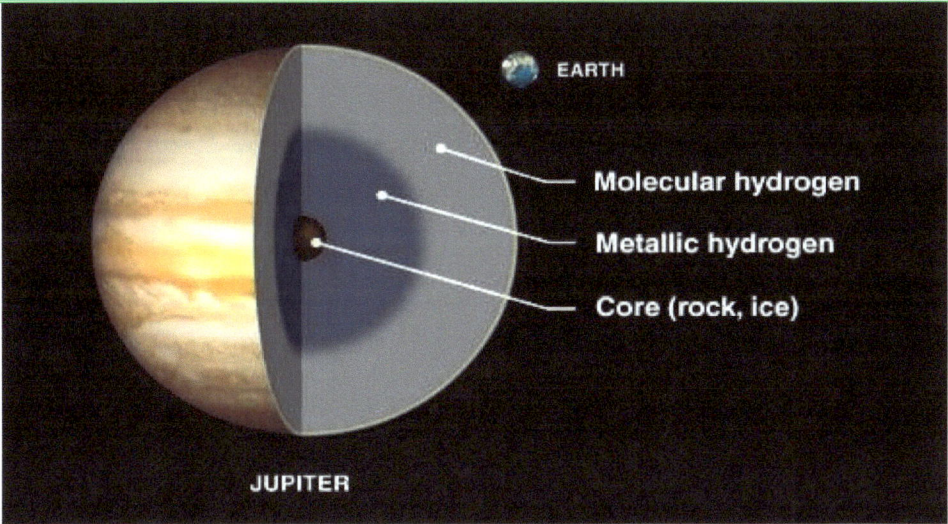

- Its diameter is 11 times bigger than that of the Earth's.
- 1300 Earths could fit inside Jupiter.
- It is sometimes called a "mini-solar system" because of its numerous moons and several rings.
- Jupiter appears striped because light and dark belts are created by strong east-west winds.
- It is the planet with the highest gravity in the solar system
- Atmosphere: Mostly hydrogen, helium and methane.

SATURN

- **6th** planet out from the sun.
- **Second** Largest Planet
- **Saturn** is composed almost entirely of hydrogen and helium.
- Has **Rings** that are made up of **frozen gas**, **ice**, and **rock**.
- **Saturn's rings** are very wide.
- They **extend outward** to about 260,000 miles from the **surface** but are less than 1 mile **thick**.

SATURN

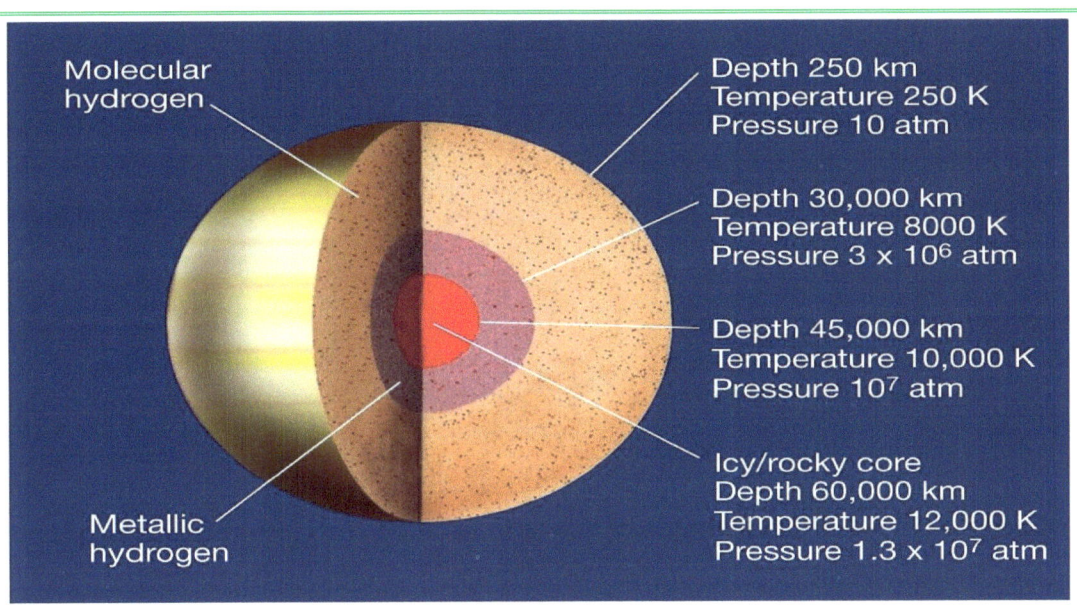

- 800 Earths could fit inside Saturn.
- It takes Saturn about 30 years to orbit the sun.
- Saturn is Yellow in Color.
- (Moons): 60 Moons; been named and others have alphanumeric designations
- Saturn has a large system of rings, and the yellow and gold bands in its atmosphere are caused by fast winds combined with heat rising from its interior.

URANUS

- **7th** planet out from the sun.
- **Third** Largest Planet.
- **Uranus** is blue in color due to methane gas in its atmosphere.
- **Uranus** has 11 dark rings surrounding it.
- **Uranus** has 21 **moons**
- **Uranus** takes 84 years to complete one orbit.

URANUS

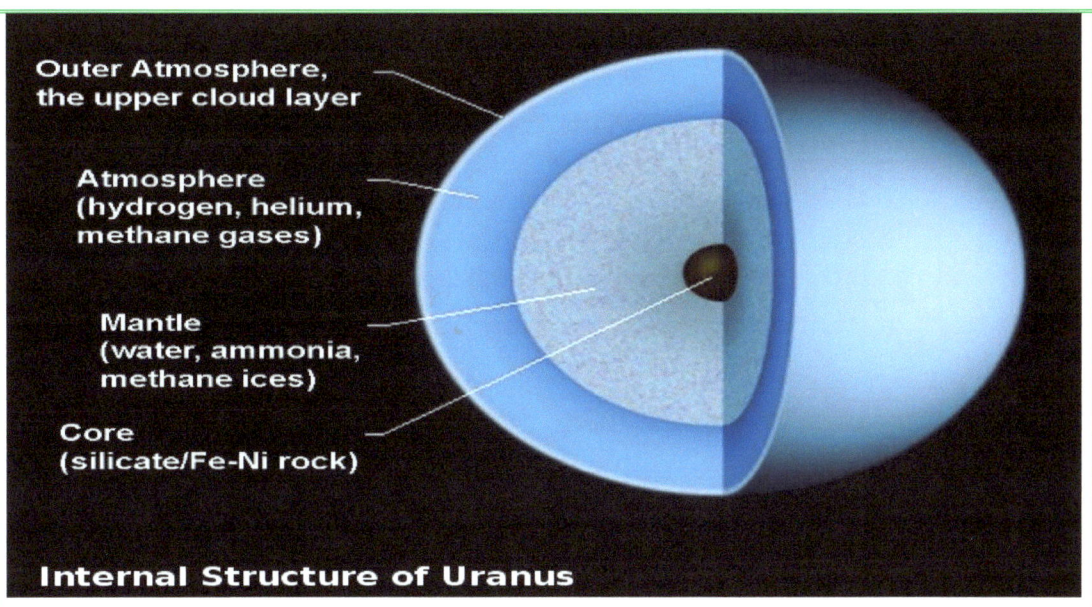

*Looks like it is on its side.

*It has a blue-green color from the methane gas above the deeper clouds.

Methane absorbs red light and reflects blue light. It does have a small system of rings.

*Four times larger in diameter than the Earth.

*Moons: 27 Moons.

NEPTUNE

- 8th planet out from the sun.
- Fourth Largest Planet.
- Size: 44 times the volume of the Earth.
- Neptune is also blue in color due to methane gas in its atmosphere.
- Neptune takes 165 years to orbit the sun and has 8 moons.
- Two dark spots.

NEPTUNE

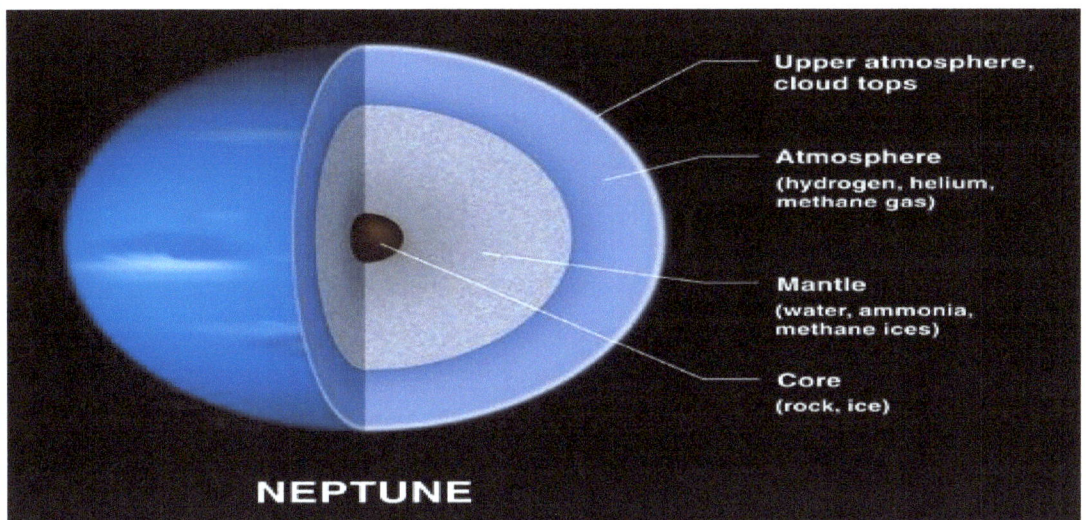

Internal structure

*Seven thin rings.	*Neptune has the fastest winds in the solar system: up to 2,000 km/hr.

*Neptune has a blue color because of the methane in its atmosphere.

*The methane reflects blue light while it absorbs red light.

*It has a small system of rings and periodically Great Dark Spots (hurricane-like storms) appear.

*Neptune has no solid surface, but its liquid core is composed of water and other "melted ice."	*Moons: 13 Moons.

ASTEROIDS

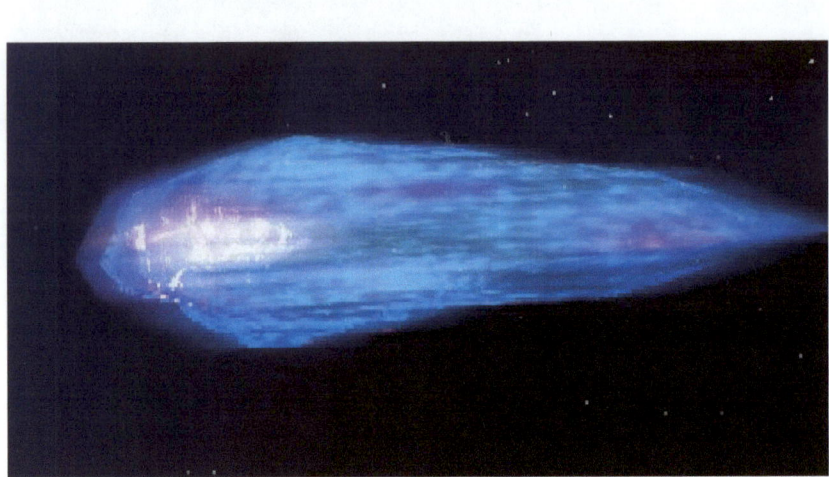

- Asteroids are either rocky or metallic objects that orbit the Sun.

- Asteroids are too small to considered planets but are sometimes called planetoids.

- They can be anywhere from the size of a pebble up to a 1000km (620 miles) in diameter; the asteroid Ceres is an example of an asteroid that is this large.

- They have been found inside Earth's orbit and all the way out past Saturn's orbit.

- Most asteroids, however, are located in the asteroid belt which exists between the orbit's of Mars and Jupiter.

ASTEROID BELT

The Main Asteroid Belt
(Orbits drawn approximately to scale)

Jupiter

Mars

Sun

Asteroid Belt

- They are known as minor planets.

COMETS

*Comets – small, fragile, irregular-shaped body composed of a mixture of non-volatile grains and frozen gases.

*Their orbits are elliptical (oval) or parabolic (U-shaped).

*The orbit brings them in very close to the Sun and swings them far out into space, sometimes out past Pluto. .

*As comets approach the Sun, radiation from the Sun evaporates the ice and gases, creating the lone tail.

*The closer to the Sun, the longer the tail of the comet.

*The tail of the comet always faces away from the Sun because of the solar wind coming from the Sun.

COMPONENTS OF COMETS

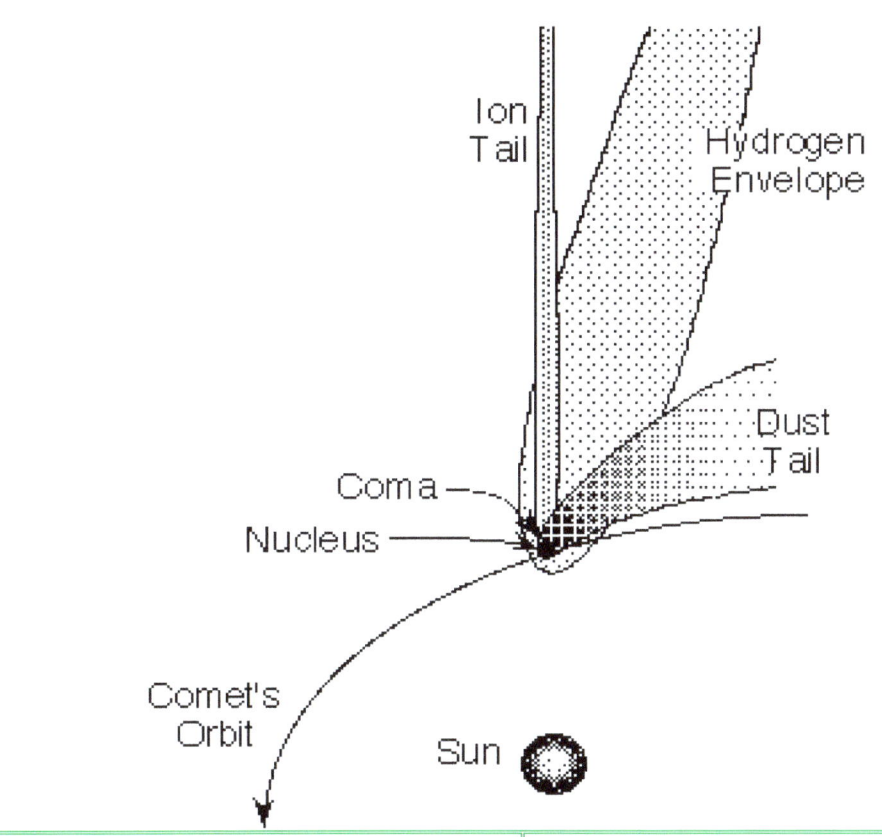

- Bodies in space made up of ice, dust, small gritty particles.
- When close to the sun, ice vaporizes, producing a spectacular streak of gas, referred to as a "tail".
- Sometimes called "dirty snowballs".
- Many in a regular orbit around the sun.

METEORS, METEOROIDS, AND METEORITES

- **Meteoroids** similar to an asteroid but significantly **smaller**.
- Most are debris asteroids/comets

- **Meteor** – flash of light we see (shooting star) when a small chunk of interplanetary debris burns up as it passes through our atmosphere.

- **Meteorite** – what is left that strikes the Earth's surface if the meteoroid does not burn up completely.

- 92.8% of all **meteorites** are composed of **silicate** (stone), and 5.7% are composed of **iron** and **nickel**; the rest are a mixture of the three materials

A DAY ON :

Mercury   =   3 Earth months
Venus   = 117 Earth days
Mars   = 41 minutes longer than an Earth day.
Jupiter   = 10 Earth hours
Saturn   = ½ Earth hours
Uranus   = 13 ½ Earth hours
Neptune   = 18 Earth hours
Pluto   = 7 Earth days

If you weigh 85 lbs. On Earth, you weigh.

32 lbs on Mercury

77 lbs on Venus

32 lbs on Mars

244 lbs on Jupiter

112 lbs on Saturn

79 lbs on Uranus

105 lbs on Neptune

2.5 lbs on Pluto

The Orbital Speed of the Planets

- Mercury = 30 miles per second
- Venus = 22 miles per second
- Earth = 19 miles per second
- Mars = 15 miles per second
- Jupiter = 8 miles per second
- Saturn = 6 miles per second
- Uranus = 4.2 miles per second
- Neptune = 3.3 miles per second
- Pluto = 2.9 miles per second

Therefore, the further away from the Sun, the slower the orbital speed.

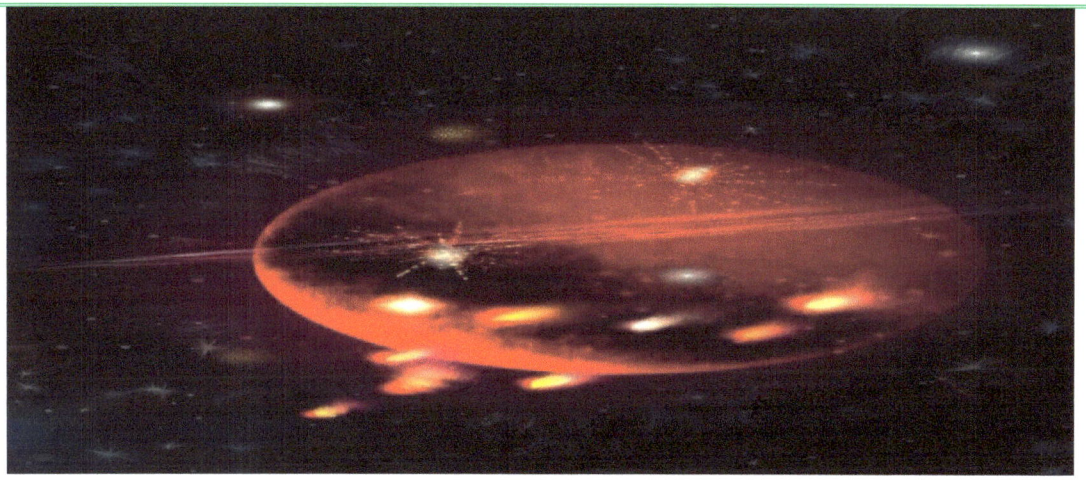

It takes **365 days** for Earth to revolve around the **Sun**.

How long does it take the other planets?

- Mercury — 88 Earth days
- Venus — 225 Earth days
- Mars — about 2 Earth years
- Jupiter — 11.8 Earth years
- Saturn — 30 Earth years
- Uranus — 84 Earth years
- Neptune — 165 Earth years
- Pluto — 248 Earth years

OTHER BOOKS BY THE AUTHOR

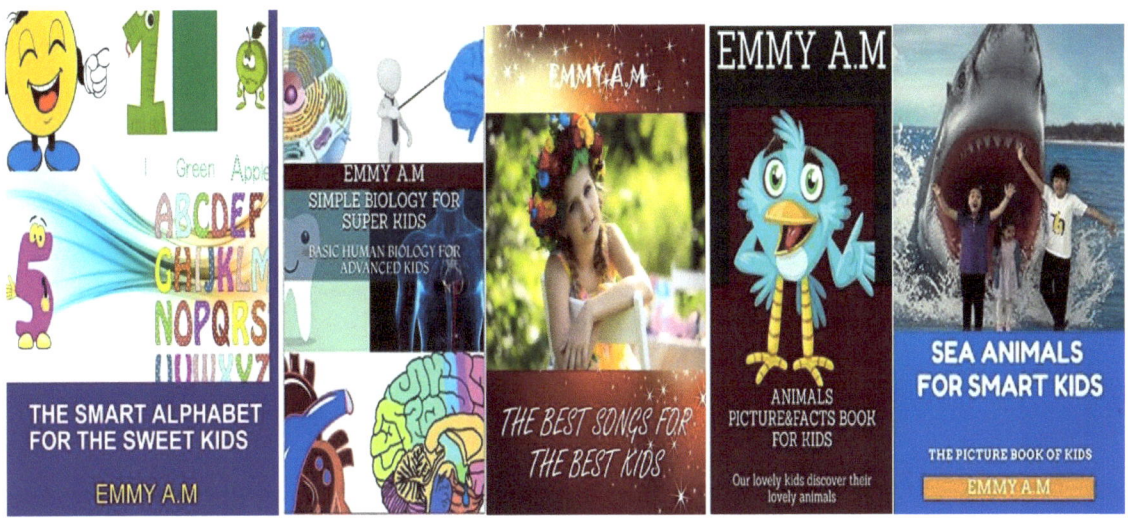

THE SMART ALPHABET FOR THE SWEET KIDS

https://www.amazon.com/dp/B07MD6PXS3

SIMPLE BIOLOGY FOR SUPER KIDS

https://www.amazon.com/dp/B07L5W42C8

THE BEST SONGS FOR THE BEST KIDS

https://www.amazon.com/dp/B07KKZ7YML

ANIMALS PICTURE&FACTS BOOK FOR KIDS

https://www.amazon.com/dp/B07KTGXM6R

SEA ANIMALS FOR SMART KIDS: THE PICTURE BOOK OF KIDS

https://www.amazon.com/dp/B07M8P6JQJ

*If you really love this book
Please review it
Your few words will make a great difference to me.*

www.ingramcontent.com/pod-product-compliance
Lightning Source LLC
Chambersburg PA
CBHW040416220526
45473CB00004B/1257